Low Cost Development of Small Water Power Sites

by Hans W. Hamm

Published by

VOLUNTEERS IN TECHNICAL ASSISTANCE
3706 Rhode Island Avenue
Mount Rainier, Maryland 20822 U.S.A.

PREFACE

During the last seven years of answering individual requests from Peace Corpsmen and other community development workers, VITA has come to realize the great need for a manual on small hydroelectric power development.

VITA is an international association of more than 5,000 scientists, engineers, businessmen and educators who volunteer their talent and spare time to help people in developing areas with their technical problems. The Volunteers are from the United States and 100 other countries.

The difficulty of communication has proved extreme in answering requests concerning the feasibility of a small hydro plant as a source of power, as compared with a diesel. The value of a manual written in simple terms is readily apparent.

The present manual has been prepared to fill this need. It should enable the reader to assess the possibility and desirability of installing a small hydroelectric power plant, select the type of machinery most suitable for installation, and order turbine and generating equipment. It should also serve as a guide in actual construction and installation. When further guidance is needed, VITA can put the reader in touch with expert VITA Volunteers.

The manual begins by describing in simple language the steps necessary to measure the head (the height of a body of water, considered as causing pressure) and flow of the water supply, and gives data for computing the amount of power available. Next it describes the construction of a small dam and points out safety precautions necessary in designing and building such structures. Following this is a discussion of turbines and water wheels. Guide lines are given for making the right choice for a particular site. In this connection, ready-made units are available from such reliable manufacturers as James Leffel & Company in the United States and Ossberger-Turbinenfabrik in Germany. Both companies give excellent service in advising prospective purchasers.

This section of the manual also describes in detail how to make a Michell (or Banki) turbine in a small machine shop with welding facilities, from usually available pipe and other stock material. However, the hazards accompanying the manufacture of so delicate a machine by do-it-yourself methods, and the difficulty of achieving high efficiency should warn the ambitious amateur to consider the obvious alternative of securing advice from a reliable manufacturer before attempting to build his own. Table 3 gives information on the availability of manufactured units. Electric generator equipment is standardized and readily available.

Appendix 1 gives detailed information on manufacturers of turbines. Appendix 2 is a chart for converting English units of measure to metric units. English units are used in the text.

Finally, for those who are interested in pursuing the subject further and who have the engineering background to understand technical treatises, a bibliography in Appendix 2 describes textbooks and handbooks available in English in the United States and England.

Harry Wiersema

TABLE OF CONTENTS

I. INTRODUCTION

A. Alternatives

Flowing water tends to generate automatically a picture of "free" power in the eyes of the observer. But there is always a cost to producing power from water sources. The cost of developing low-output water power sites should be checked against available alternatives, such as:

1. Electric Utility - wherever transmission lines can furnish unlimited amounts of reasonably priced electric current, it is usually uneconomical to develop small and medium-sized sites.

2. Generators - diesel engines and internal-combustion engines may use a variety of fuels, for example, oil, gasoline, or wood. In general, the capital expenditure for this type of power plant is low compared to a hydro-electric plant. Operating costs, on the other hand, are very low for hydro-electricity and high for generated power.

3. Solar Heat - extensive experimental work has been done on the utilization of solar heat. Equipment now available may be less costly than water power development in regions with long hours of intense sunshine.

B. Evaluation

For isolated communities in countries where the cost of coal and oil is high and access to transmission lines is limited or non-existent, development of even the smallest water power site may be worthwhile. Particularly favorable is the situation where the head (the height of a body of water, considered as causing pressure) is relatively high, and for this reason a fairly inexpensive turbine can be used (note Figure 1). Water power is also very economical where a dam can be built into a small river with a relatively short (less than 100 feet)[1] conduit (penstock) for conducting water to the water wheel (note Figure 10). Development cost can be fairly high when such a dam and pipeline can provide a head of only 20 feet or less. Cost factors which must be considered are:

1. Capital Expenditures

   a. Design cost - can be relatively high for small plants.

   b. Cost of Real Estate -
   High for low-head plants where a dam and reservoir have to be created.
   Small for high-head plants with only an intake, a pipeline and shed for machinery.

[1]A table for converting English units to metric units is given in Appendix 2.

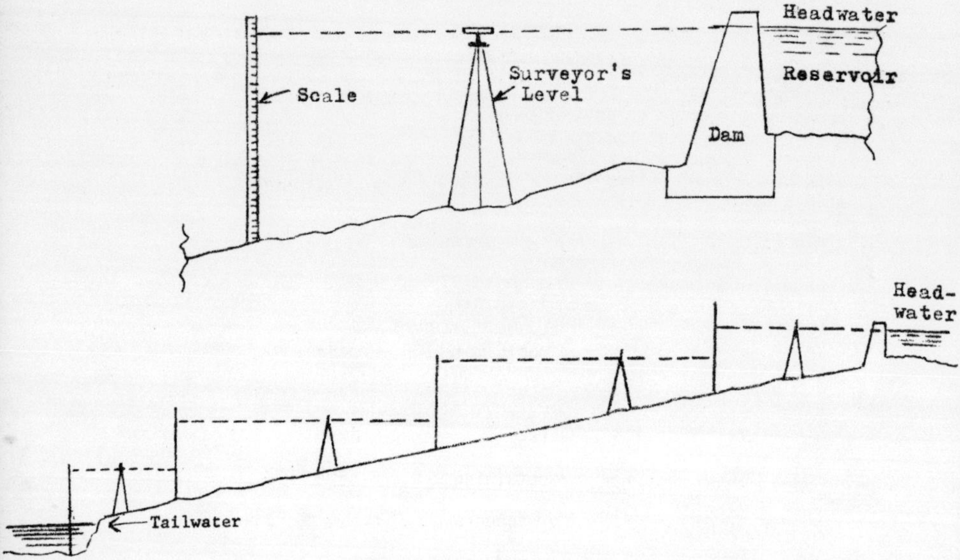

Figure 1. Measuring head with surveyor's level
(Total Gross Head = A + B + C + D)

c. Riparian Rights - the rights of those whose property borders on a body of water must be respected.
d. Construction Cost - including civil works and machinery.
e. Electrical Equipment - transformers, transmission lines, and meters.

2. Operating Expenditures

a. Amortization charges and interest on capital expenditures.
b. Depreciation - for machinery, about 4% a year.
- for buildings, it can be as low as 1% a year.
c. Labor - operation and maintenance.
d. Repairs.
e. Taxes, insurance, and administration.

The safest method of evaluating and developing a small site is to be guided by the following instructions for determining available head, flow, and, therefore, power.

A Note of Caution: flow should be measured at a time when it is at a minimum, i.e., during the dry season. Otherwise the plant will be oversized.

The data obtained can be submitted through VITA to several manufacturers of small turbines for preliminary quotations and recommendations. Turbine manufacturers will furnish considerable advice and usually an outline drawing of the entire project. Government publications for designing civil works such as a dam are available from:

U.S. Government Printing Office
Washington, D.C. 20402
U.S.A.

and

Her Majesty's Stationery Office
London, England

These agencies will supply a list of publications on the subject.

II. BASIC DATA

A. Minimum flow in cubic feet or cubic meters per second.

B. Maximum flow to be utilized.

C. Available head in feet or meters.

D. Pipe line length required for obtaining desired head.

E. Site sketch with elevations, or topographical map with site sketched in.

F. Water condition, whether clear, muddy, sandy, acid, etc.

F. Soil condition, the velocity of the water and the size of the ditch or channel for carrying it to the works depends on soil condition.

H. Minimum tailwater elevation at the powerhouse site must be given to determine the turbine setting and type.

I. Air temperature, minimum and maximum.

III. POWER

The amount of power desired (useful power) should be determined in advance. Power may be expressed in terms of horsepower or kilowatts. One horsepower is equal to 0.7455 kilowatts. One kilowatt is about one and a third horsepower. The required amount of power (gross power) is equal to the useful power plus the losses inherent in any power scheme. It is usually safe to assume that the net or useful power in the case of small power installations will only be half of the available gross power due to water transmission losses and the turbine and generator efficiencies. Some power is lost when it is transmitted from the generator switchboard to the place of application.

The GROSS POWER, the power available from the water, is determined by the following formula:

In English Units:

Gross Power (horsepower) =

$$\frac{\text{Minimum Water Flow (cubic feet/second) X Gross Head(feet)}}{8.8}$$

In Metric Units:

$$\text{Gross Power (Metric horsepower)} = \frac{1{,}000}{75}\ \text{Flow (cubic meters/second) X Head(meters)}$$

The NET POWER available at the turbine shaft is:

In English Units:

$$\text{Net Power} = \frac{\text{Minimum Water Flow X Net Head}}{8.8}\ \text{X Turbine Efficiency (English)}$$

In Metric Units:

$$\text{Net Power} = \frac{\text{Minimum Water Flow X Net Head}}{75/1{,}000}\ \text{X Turbine Efficiency (Metric)}$$

The NET HEAD is obtained by deducting the energy losses from the gross head. These losses are discussed in section VI. A good assumption for turbine efficiency, when it is not known, is 80%.

IV. MEASURING GROSS HEAD
(Either Method)

A. Method No. 1

1. Equipment
   a. Surveyor's leveling instrument - consists of a spirit level fastened parallel to a telescopic sight (note Figure 2).
   b. Scale - use wooden board approximately 12 feet in length (note Figure 3).

2. Procedure (note Figure 1)
   a. Surveyor's level on a tripod is placed down stream from the power reservoir dam on which the headwater level is marked.
   b. After taking a reading, the level is turned 180° in a horizontal circle. The scale is placed downstream from it at a suitable distance and a second reading is taken. This process is repeated until the tailwater level is reached.

Figure 2. Surveyor's Level

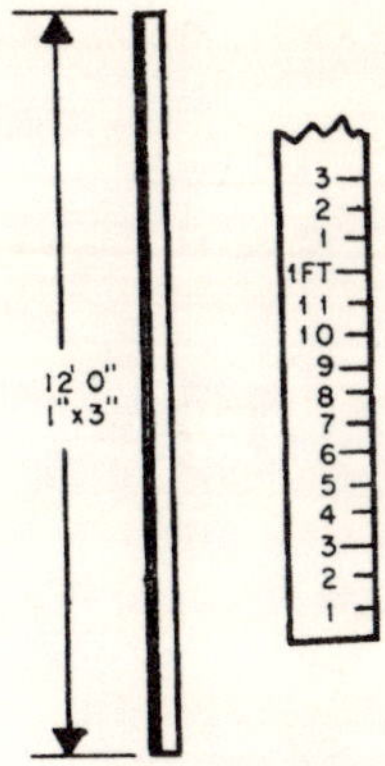

Figure 3. Scale and Detail of Scale

B. <u>Method No. 2</u>

This method is fully reliable, but is more tedious than Method No. 1 and need only be used when a surveyor's level is not available.

1. Equipment
   a. Scale (note Figure 3).
   b. Board and wooden plug (note Figures 4 and 6).
   c. Ordinary carpenter's level (note Figure 5).

2. Procedure (note Figure 6)
   a. Place board horizontally at headwater level and place level on top of it for accurate leveling. At the downstream end of the horizontal board, the distance to a wooden plug set into the ground is measured with a scale.
   b. The process is repeated step wise until the tailwater level is reached.

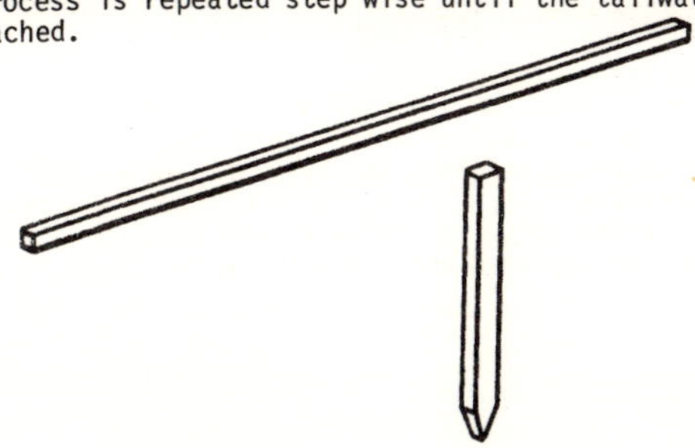

Figure 4. Leveling Board and Plugs in Stokes

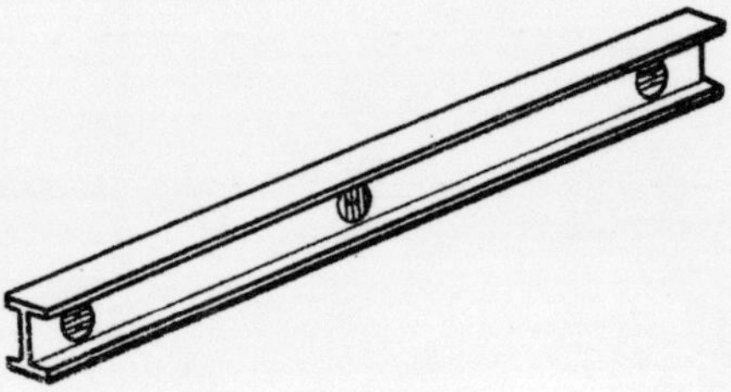

Figure 5. Carpenter's Level

V. MEASURING FLOW RATE

For power purposes, measurements should take place at the season of lowest flow in order to guarantee full power at all times. Investigate the stream flow history to ascertain that the minimum required flow is that which has occurred for as many years as it is possible to determine. An obvious point that has, nevertheless, been overlooked in the past is this: if there have been years of drought in which flow rate was reduced below the minimum required, other streams or sources of power may offer a better solution.

A. Method No. 1

For small streams with a capacity of less than one cubic foot per second, build a temporary dam in the stream, or use a "swimming hole" created by a natural dam. Channel the water into a pipe and catch it in a bucket of known capacity. Determine the stream flow by measuring the time it takes to fill the bucket.

$$\text{Stream Flow (cubic feet per second)} = \frac{\text{Volume of Bucket (cubic feet)}}{\text{Filling Time (seconds)}}$$

B. Method No. 2

For medium streams with a capacity of more than one cubic foot per second, the weir method can be used. The weir (see Figures 7 & 8) is made from boards, logs or scrap lumber. Cut a rectangular opening in the center. Seal the seams of the boards and the sides built into the banks with clay or sod to prevent leakage. Saw the edges of the opening on a slant to produce sharp edges on the upstream side. A small pond is formed upstream from the weir. When

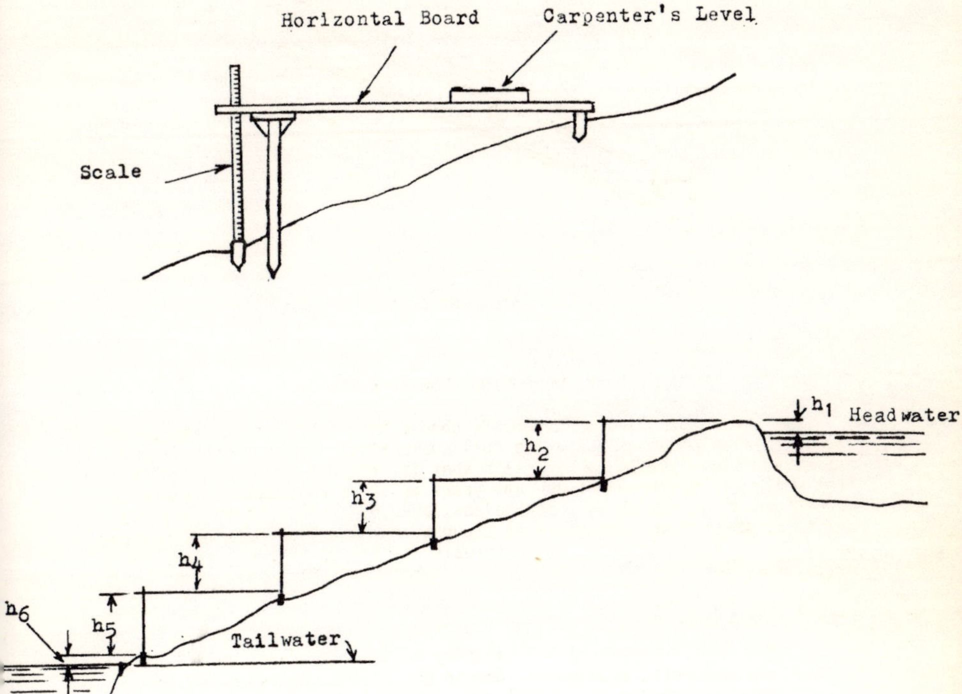

$$\text{Total Gross Head } H_g = h_2 + h_3 + h_4 + h_5 + h_6 - h_1$$

Figure 6. Measuring head with carpenter's level

there is no leakage and all water is flowing through the weir opening, (1) place a board across the stream and (2) place another narrow board level (use a carpenter's level) and perpendicular to the first. Measure the depth of the water above the bottom edge of the weir with the help of a stick on which a scale has been marked. Determine the flow from Table I.

Figure 7. The weir method of measuring flow

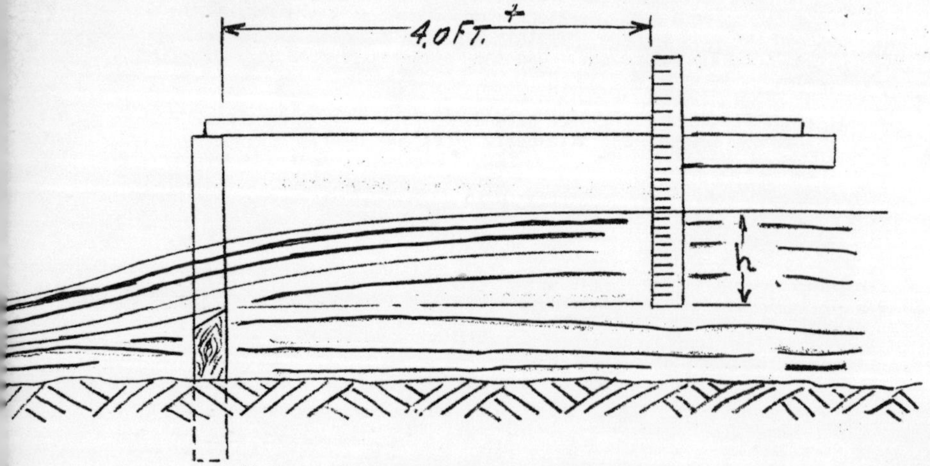

Figure 8. Measuring flow with a weir

Table I

FLOW VALUE (Cubic Feet per Second)

Overflow Height	Weir Width 3 feet	4 feet	5 feet	6 feet	7 feet	8 feet	9 feet
1.0 inch	.24	.32	.40	.48	.56	.64	.72
2 inches	.67	.89	1.06	1.34	1.56	1.8	2.0
4 inches	1.9	2.5	3.2	3.8	4.5	5.0	5.7
6 inches	3.5	4.7	5.9	7.0	8.2	9.4	10.5
8 inches	5.4	7.3	9.0	10.8	12.4	14.6	16.2
10 inches	7.6	10.0	12.7	15.2	17.7	20.0	22.8
12 inches	10.0	13.3	16.7	20.0	23.3	26.6	30.0

C. Method No. 3

The float method (Figure 9) is used for larger streams. Although it is not as accurate as the previous two methods, it is adequate for practical purposes. Choose a point in the stream where the bed is smooth and the cross section is fairly uniform for a length of at least 30 feet. Measure water velocity by throwing pieces of wood into the water and measuring the time of travel between two fixed points, 30 feet or more apart. Erect posts on each bank at these points. Connect the 2 upstream posts by a level wire rope (use a carpenter's level). Follow the same procedure with the down stream posts. Divide the stream into equal sections along the wires and measure the water depth for each section. In this way, the cross-sectional area of the stream is determined. Use the following formula to calculate the flow:

Stream Flow (cubic feet per second) = Average Cross-Sectional Flow Area(square feet) X Velocity (feet per second)

VI. MEASURING HEAD LOSSES

As noted in Section III, the "Net Power" is a function of the "Net Head". The "Net Head" is the "Gross Head" less the "Head Losses". Figure 10 shows a typical small water power installation. The head losses are the open-channel losses plus the friction loss from flow through the penstock.

A. Open Channel Head Losses

The headrace and the tailrace in Figure 11 are open channels for transporting water at low velocities. The walls of channels, made of timber, masonry, concrete, or rock, should be constructed perpendicularly. Design them so that the water level height is one half of the width. Earth walls should be built at a 45° angle. Design them so that the water level height is one half of the channel width at the bottom. At the water level the width is twice that of the bottom.

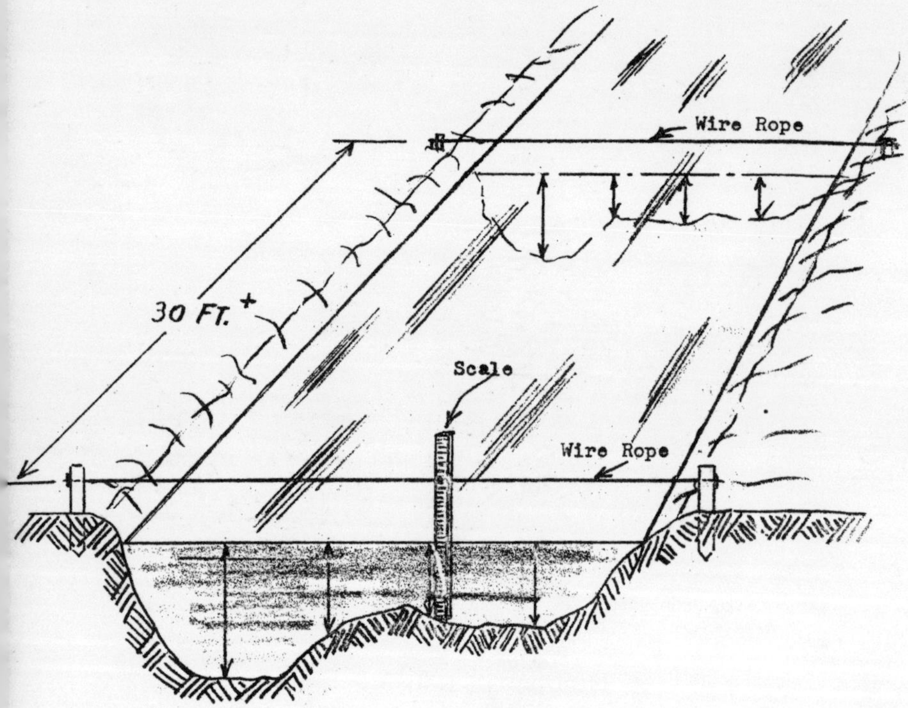

Figure 9. The float method of measuring flow

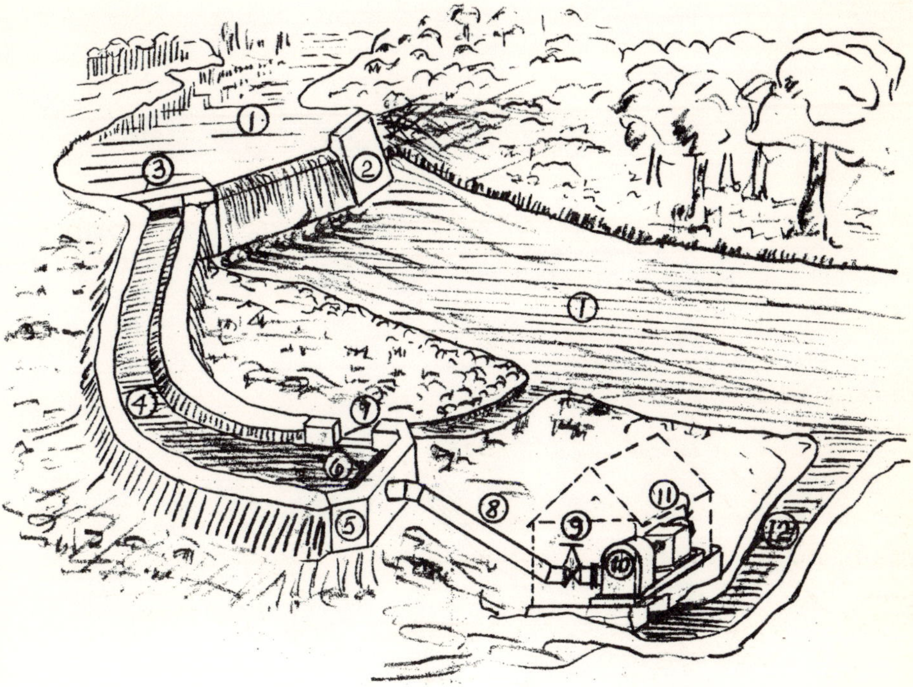

Figure 10. A typical installation for a low-output water power plant

1. River
2. Dam with Spillway
3. Intake to Headrace
4. Headrace
5. Intake to Turbine Penstock
6. Trashrack
7. Overflow of Headrace
8. Penstock
9. Turbine Inlet Valve
10. Water Turbine
11. Electric Generator
12. Tailrace

For Timber, Concrete, Masonry & Rock

Hydraulic Radius = 0.25 W

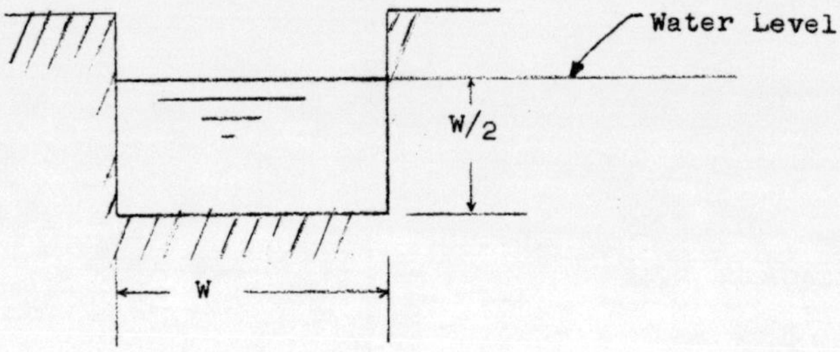

W = Bottom Width

For Earth Channels

Hydraulic Radius = 0.31 W

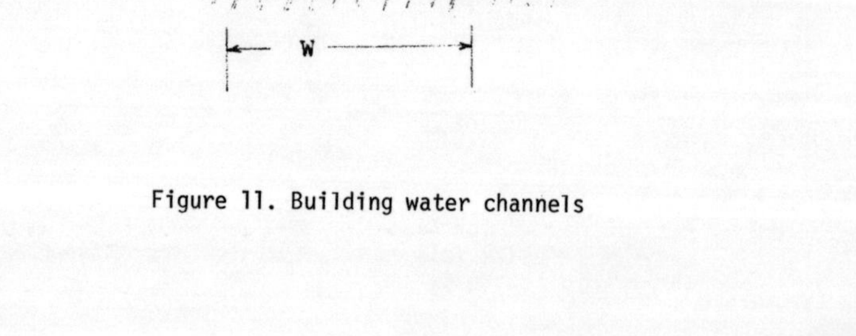

Figure 11. Building water channels

The head loss in open channels is given in the nomograph in Figure 12. The friction effect of the material of construction is called "n". Various values of "n" and the maximum water velocity, below which the walls of a channel will not erode are given in Table II.

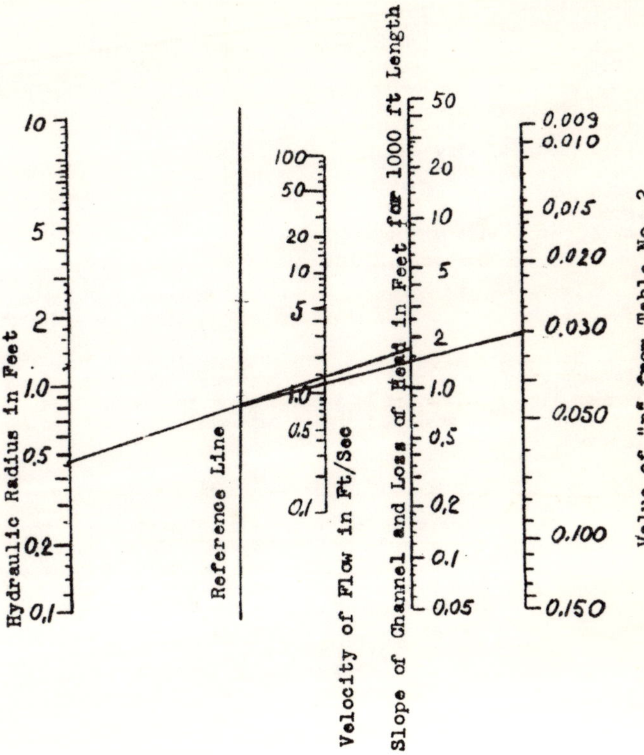

Figure 12. Fall of Channel (or Slope) in feet per 1000 ft. of Channel Length (The total fall is equal to the Loss of Head in Feet through the Channel)

Table II

Material of Channel Wall	Maximum Allowable Water Velocity (feet/second)		Value of "n"	
Fine grained sand	0.6		0.030	
Coarse sand	1.2		0.030	
Small stones	2.4		0.030	
Coarse stones	4.0		0.030	
Rock	25.0	(Smooth)	0.033	(Jagged) 0.045
Concrete with sandy water	10.0		0.016	
Concrete with clean water	20.0		0.016	
Sandy loam, 40% clay	1.8		0.030	
Loamy soil, 65% clay	3.0		0.030	
Clay loam, 85% clay	4.8		0.030	
Soil loam, 95% clay	6.2		0.030	
100% clay	7.3		0.030	
Wood			0.015	
Earth bottom with rubble sides			0.033	

The hydraulic radius is equal to a quarter of the channel width, except for earth-walled channels where it is 0.31 times the width at the bottom.

To use the nomograph, a straight line is drawn from the value of "n" through the flow velocity to the reference line. The point on the reference line is connected to the hydraulic radius and this line is extended to the head-loss scale which also determines the required slope of the channel.

B. Pipe Head loss and Penstock Intake

The trashrack in Figure 13 is a weldment consisting of a number of vertical bars held together by an angle at the top and a bar at the bottom. The vertical bars must be spaced in such a way that the teeth of a rake can penetrate the rack for removing leaves, grass and trash which might clog up the intake. Such a trashrack can easily be manufactured in the field or in a small welding shop. Downstream from the trashrack, a slot is provided in the concrete into which a timber gate can be inserted for shutting off the flow of water to the turbine.

The penstock can be constructed from commercial pipe. The pipe must be large enough to keep the head loss small. From the nomograph (Figure 14) the required pipe size is determined. A straight line drawn through the water velocity and flow rate scales gives the required pipe size and pipe head-loss. Head loss is given for a 100-foot pipe length. For longer or shorter penstocks, the actual head loss is the head loss from the chart multiplied by the actual length divided by 100. If commercial pipe is too expensive, it is possible to make pipe from native material; for example, concrete and ceramic pipe or hollowed logs. The choice of pipe material and the method of making the pipe depend on the cost and availability of labor and the availability of material. VITA can provide the needed technical information.

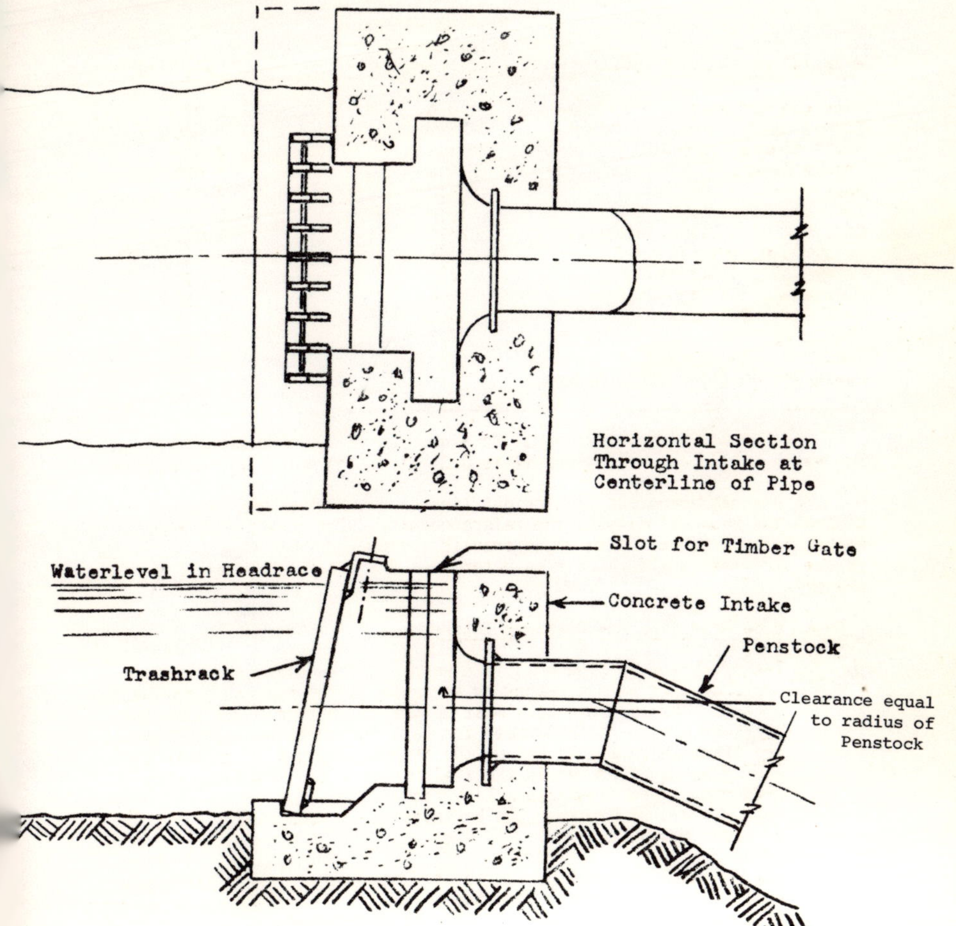

Figure 13. Sectional Elevation: Intake to penstock of small water turbine

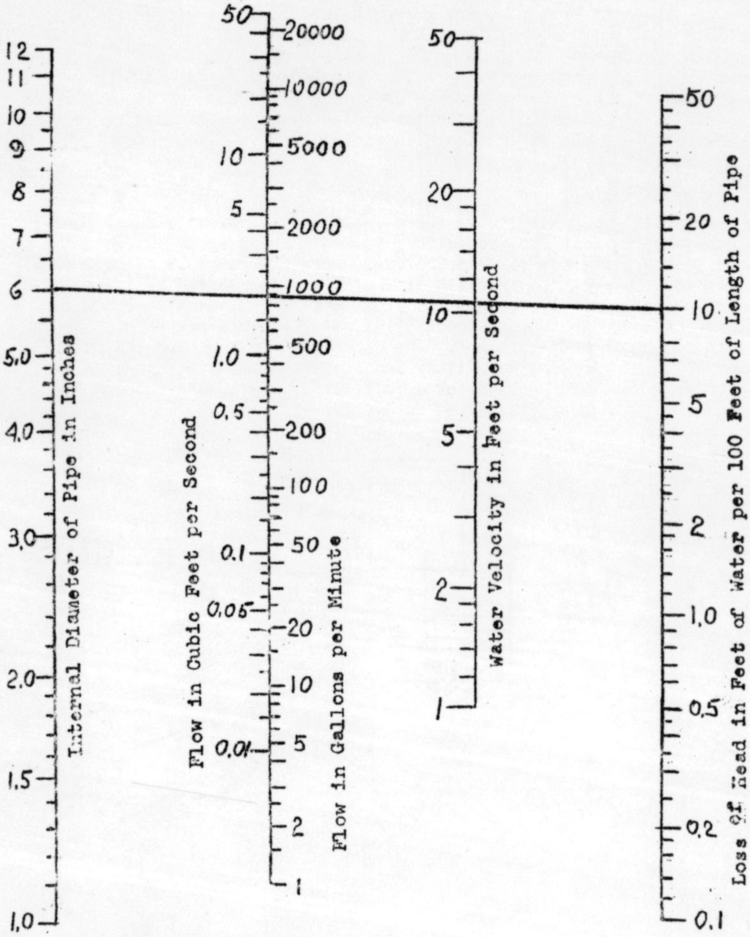

Figure 14. Nomograph for head loss in steel pipes

VII. SMALL DAMS

A dam is necessary in most cases to direct the water into the channel intake or to get a higher head than the stream naturally affords. A dam is not required if there is enough water to cover the intake of a pipe or channel at the head of the stream where the dam would be placed.

A dam may be made of earth, wood, concrete or stone. In building any kind of a dam, all mud, vegetable matter and loose material must be removed from the bed of the stream where the dam is to be placed. This usually is not difficult since most small streams will cut their beds down close to bed rock, hard clay or other stable formation.

A. Earth Dams

An earth dam may be desirable where concrete is expensive and timber scarce. It must be provided with a separate spillway of sufficient size to carry off excess water because water can never be allowed to flow over the crest of an earth dam. If it does the dam will erode and be destroyed. A spillway must be lined with boards or with concrete to prevent seepage and erosion. Still water is held satisfactorily by earth but moving water is not. The earth will be worn away by it. Figures 15 and 16 show a spillway and an earth dam. The crest of the dam may be just wide enough for a footpath or may be wide enough for a roadway, with a bridge placed across the spillway.

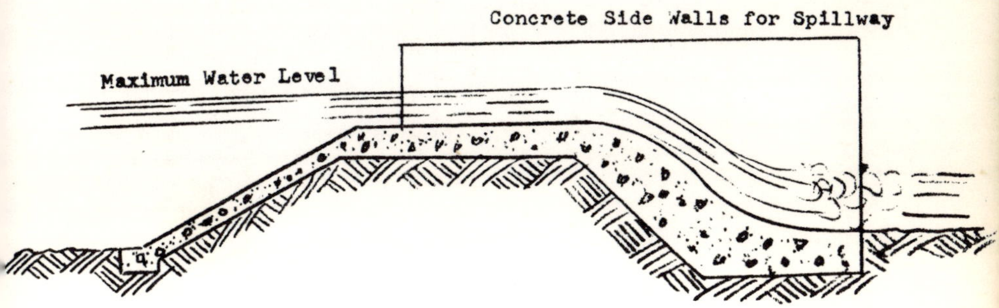

Figure 15. Concrete spillway for earth-fill dam

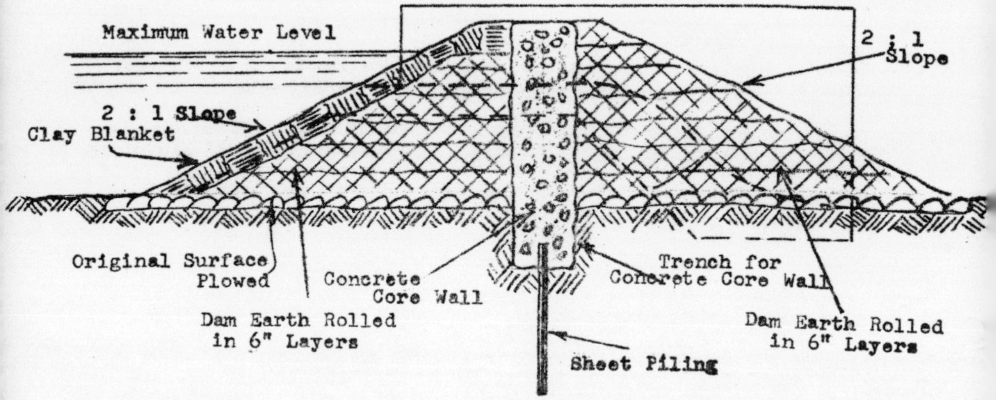

Figure 16. Earth-fill dam

The greatest difficulty in earth-dam construction occurs in places where the dam rests on solid rock. It is hard to keep the water from seeping between the dam and the earth and finally undermining the dam. One way of preventing seepage is to blast and clean out a series of ditches in the rock, with each ditch about a foot deep and two feet wide extending under the length of the dam. Each ditch should be filled with three or four inches of wet clay compacted by stamping it. More layers of wet clay can then be added and the compacting process repeated each time until the clay is several inches higher than bedrock. The upstream half of the dam, as shown in Figure 16 should be of clay or heavy clay soil, which compacts well and is impervious to water. The downstream side should consist of lighter and more porous soil, which drains out quickly and thus makes the dam more stable than if it were made entirely of clay.

B. Crib Dams

The crib dam is very economical in timber country as it requires only rough tree trunks, cut planking and stones. Four- to six-inch tree trunks are placed two to three feet apart and spiked to others placed across them at right angles. Stones fill the spaces between timbers. The upstream side (face) of the dam, and sometimes the downstream side, is covered with planks (see Figure 17). The face is sealed with clay to prevent leakage. Downstream planks are used as an apron to guide the water which overflows the dam back into the stream bed. The dam itself serves as a spillway in this case. The water coming over the apron falls rapidly and it is necessary to line the bed below with stones in order to prevent erosion. A section of a crib dam without downstream planking is illustrated in Figure 18. The apron consists of a series of steps for slowing the water gradually.

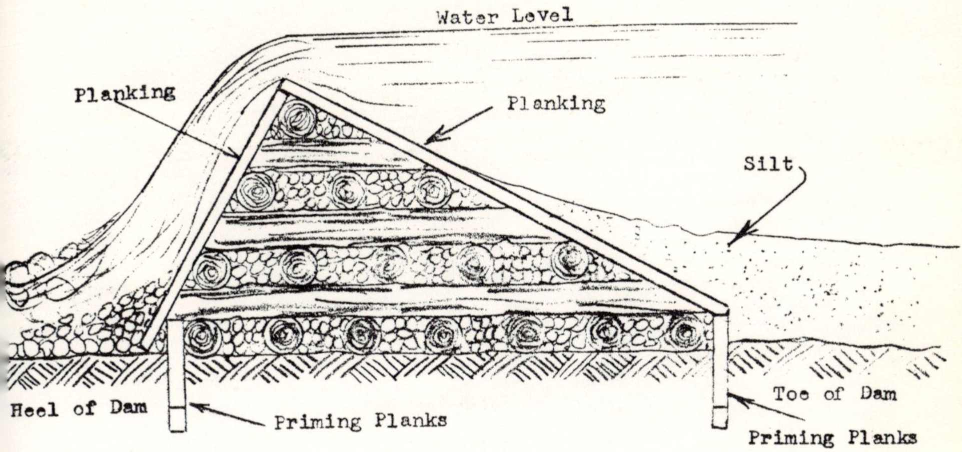

Figure 17. Crib dam with downstream planking

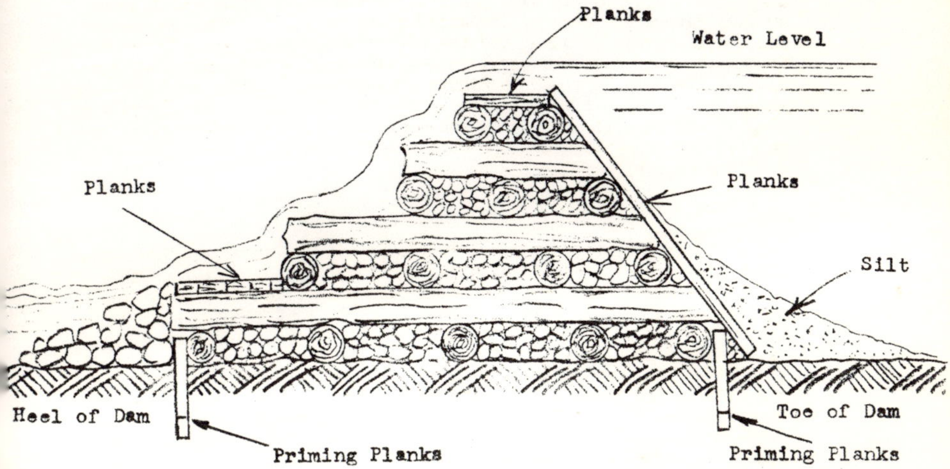

Figure 18. Crib dam without downstream planking

Crib dams, as well as other types, must be embedded well into the embankments and packed with impervious material such as clay or heavy earth and stones in order to anchor them and to prevent leakage. At the heel as well as at the toe of crib dams, longitudinal rows of planks are driven into the stream bed. These are priming planks which prevent water from seeping under the dam, and the also anchor it. If the dam rests on rock, priming planks cannot and need not be driven; but where the dam does not rest on rock they make it more stable and watertight. These priming planks should be driven as deep as possible and then spiked to the timber of the crib dam. The lower ends of the priming planks are pointed as shown in Figure 19, and they must be placed one after the other as shown. Thus each successive plank is forced, by the act of driving it, closer against the preceding plank resulting in a solid wall. Any rough lumber may be used. Chestnut and oak are considered to be the best material. The lumber must be free from sap, and its size should be approximately two inches by six inches. In order to drive the priming planks and also the sheet piling of Figure 16, considerable force may be required. A simple pile driver as shown in Figure 20 will serve the purpose.

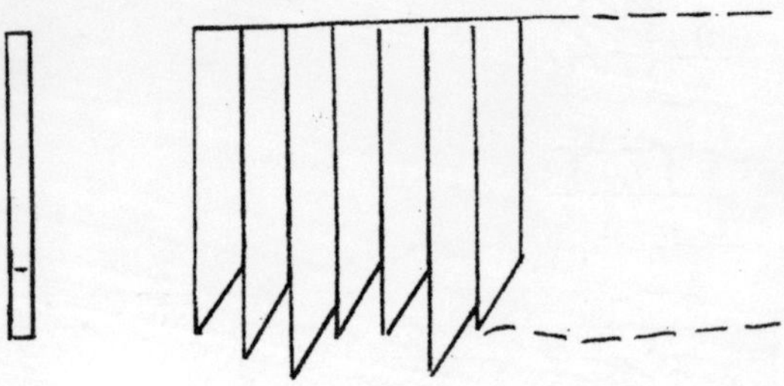

Figure 19. Priming planks

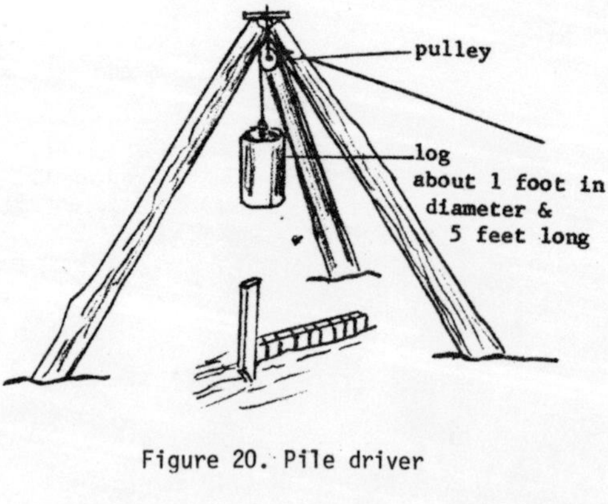

Figure 20. Pile driver

C. Concrete and Masonry Dams

Concrete and masonry dams more than 12 feet high should not be built without the advice of a competent engineer with experience in this special field. Dams of less height require knowledge of the soil condition and bearing capacity as well as of the structure itself. Figure 21 shows a stone dam which also serves as a spillway. It can be up to ten feet in height. It is made of rough stones. The layers should be bound by concrete. The dam must be built down to a solid and permanent footing to prevent leakage and shifting. The base of the dam should have the same dimension as its height to give it stability.

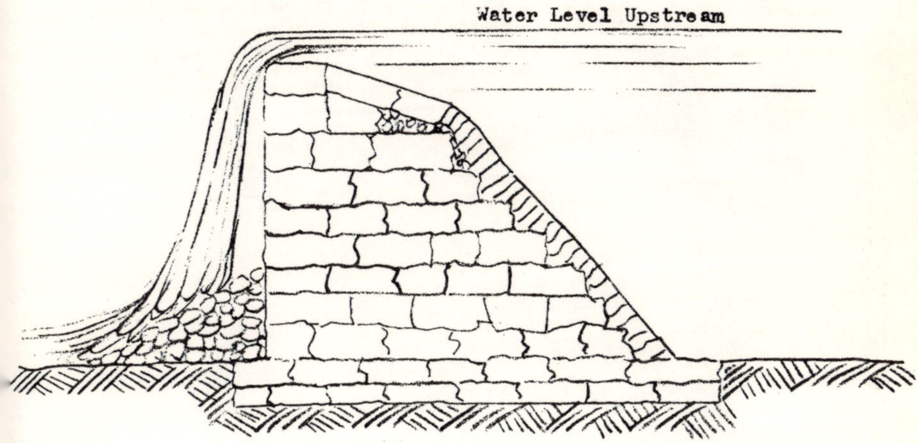

Figure 21. Stone dam

Small concrete dams (Figure 22) should have a base with a thickness 50% greater than height. The apron is designed to turn the flow slightly upwards to dissipate the energy of the water and protect the downstream bed from eroding.

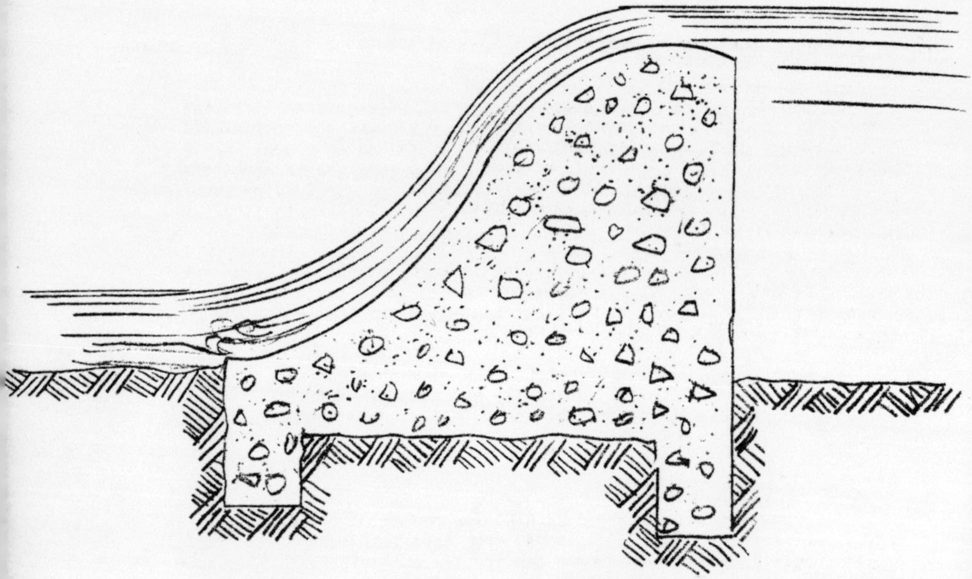

Figure 22. Small concrete dam

VIII. WATER TURBINES

The manufacturers of hydraulic turbines for small plants can usually quote on a complete packaged unit, including the generator, governor and switch gear. Water turbines for small power developments may be purchased (see Table III) or made in the field, if a small machine and weld shop is available.

A centrifugal pump may be used as a turbine wherever it is technically possible. Its cost is about one-third the cost of a hydraulic turbine. But it may be poor economics to use a centrifugal pump because it is less efficient than a turbine and will have other disadvantages.

A water power unit can produce either direct current (D.C.) or alternating current (A.C.) electricity.

Two factors to consider in deciding whether to install an A.C. or D.C. power unit are (1) the cost of regulating the flow of water into the turbine for A.C. and (2) the cost of converting motors to use D.C. electricity.

Flow Regulation

The demand for power will vary from time to time during the day. With a constant flow of water into the turbine, the power output will sometimes be greater than the demand for power. Therefore, either excess power must be stored or the flow of water into the turbine must be regulated according to the demand for power.

In producing A.C., the flow of water must be regulated because A.C. cannot be stored. Flow regulation requires governors and complex valve-type shut-off devices. This equipment is expensive; in a small water power site, the regulating equipment would cost more than a turbine and generator combined. Furthermore, the equipment for any turbine used for A.C. must be built by experienced water-turbine manufacturers and serviced by competent consulting engineers.

The flow of water to a D.C. producing turbine, however, does not have to be regulated. Excess power can be stored in a storage battery. Direct-current generators and storage batteries are low in cost because they are mass-produced.

To summarize: In producing A.C., the flow of water into the turbine must be regulated; this requires costly and complex equipment. In producing D.C., regulation is not necessary, but storage batteries must be used.

Converting Motors for D.C.

D.C. power is just as good as A.C. for producing electric light and heat. But for electrical appliances, from farm machinery to household appliances, the use of D.C. power can involve some expense. When such appliances have A.C. motors, D.C. motors must be installed. The cost of doing this must be weighed against the cost of flow regulation needed for producing A.C.

Table III

Small Hydraulic Turbines

	Types		
	Impulse or Pelton	Michell or Banki	Centrifugal Pump Used as Turbine
Head Range (feet)	50 to 1000	3 to 650	Available for any desired condition
Flow Range (cubic feet per second)	0.1 to 10	0.5 to 250	
Application	high head	medium head	
Power (horsepower)	1 to 500	1 to 1000	
Cost per Kilowatt	low	low	low
Manufacturers	James Leffel & Co. Springfield, Ohio U.S.A. 45501	Ossberger-Turbinenfabrik 8832 Weissenburg Bayern, Germany	Any reputable dealer or manufacturer
	Drees & Co. Werl, Germany	Can be do-it-yourself project if small weld and machine shops are available	
	Officine Buhler Taverne, Switzerland		

A. Impulse Turbines

Impulse turbines are used for high heads and low flow rates. They are the most economical turbine because the high head gives them high speed and their size and weight per horsepower is small. Construction costs of intake and power house are also small. A very simplified version is shown in Figures 23 and 24.

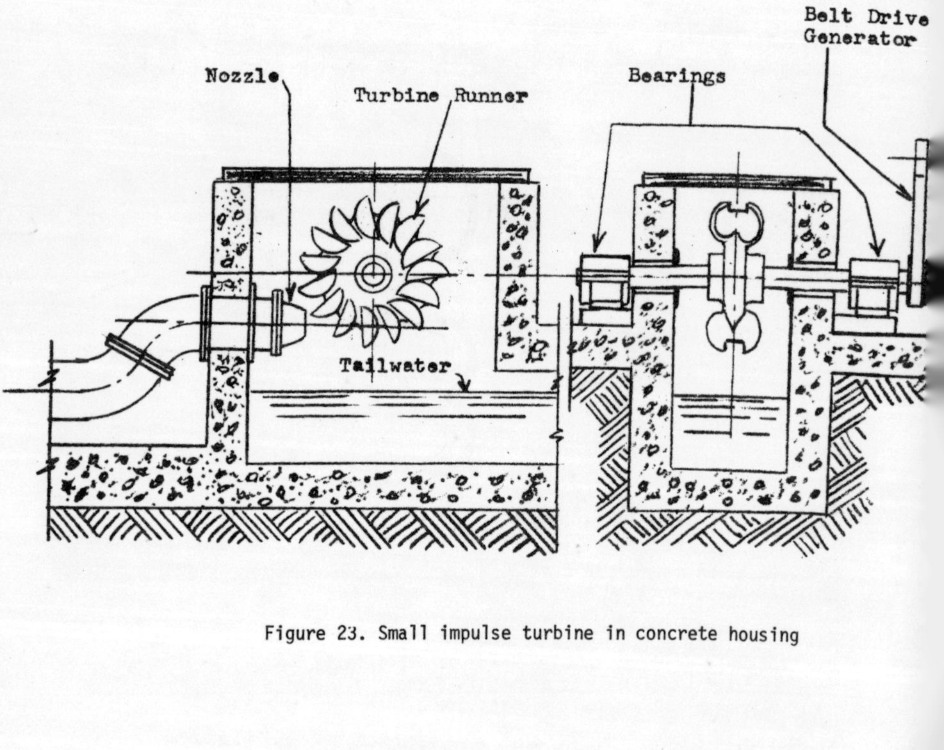

Figure 23. Small impulse turbine in concrete housing

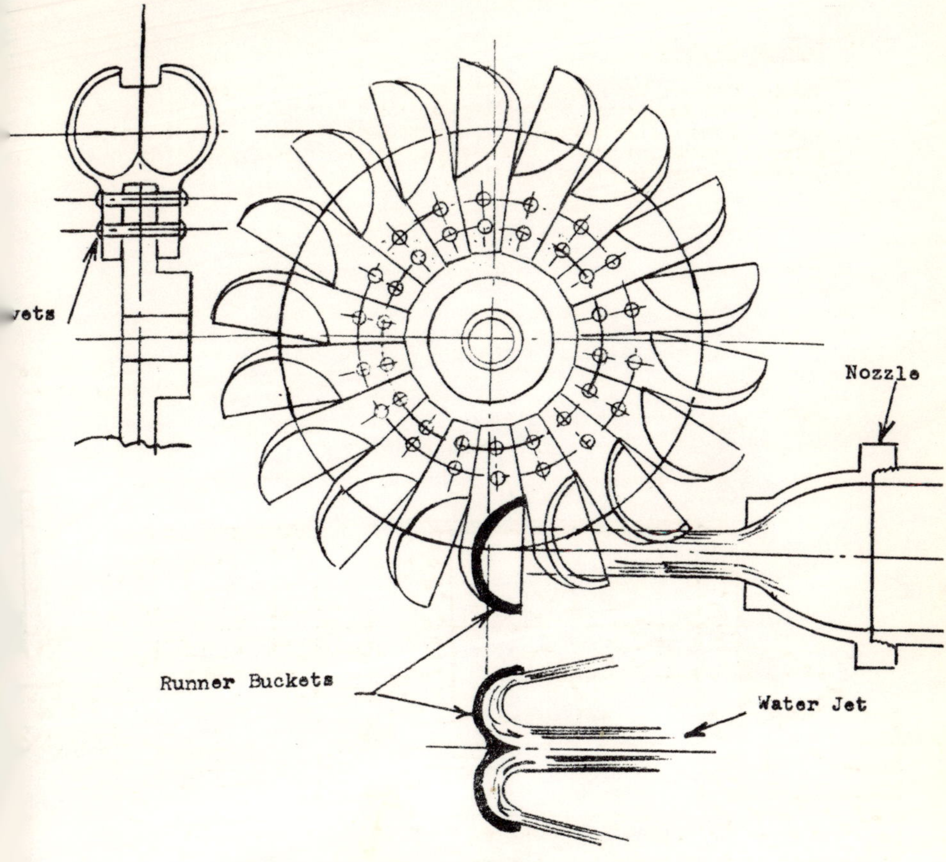

Figure 24. Small impulse-turbine runner

B. Michell (Banki) Turbine

The Michell (or Banki) turbine is simple in construction and may be the only type of water turbine which can be locally built. Welding equipment and a small machine shop like those often used to repair farm machinery and automotive parts are all that is necessary.

The two main parts of the Michell turbine are the runner and the nozzle. Both are welded from plate steel and require some machining.

Figures 25 and 26 show the arrangement of a turbine of this type for low-head use without control. This installation drives a direct-current generator with a belt drive. Because the construction can be a DO-IT-YOURSELF project, formulas and design details are given for a runner of 12" outside diameter. This size is the smallest which is easy to fabricate and weld. It has a wide range of application for all small power developments with head and flow suitable for the Michell turbine. Different heads result in different rotational speeds. The proper belt-drive ratio gives the correct generator speed. Various amounts of water determine the width of the nozzle (B_1, Figure 26) and the width of the runner (B_2, Figure 26). These widths may vary from 2 inches to 14 inches. No other turbine is adaptable to as large a range of flow.

The water passes through the runner twice in a narrow jet before discharge into the tailrace. The runner consists of two side plates, each 1/4" thick with hubs for the shaft attached by welding, and from 20 to 24 blades. Each blade is 0.237" thick and cut from 4" standard pipe. Steel pipe of this type is available virtually everywhere. A pipe of suitable length produces four blades. Each blade is a circular segment with a center angle of 72 degrees. The runner design, with dimensions for a foot-long runner, is shown in Figure 27; and Figure 28 gives the nozzle design and dimensions. The dimensions can be altered proportionally for other size runners. Upstream from the nozzle discharge opening of 1 1/4", the shape of the nozzle can be made to suit penstock pipe conditions.

To calculate the principal turbine dimension:

$$B_1 = \text{Nozzle Width (inches)} = \frac{210 \times \text{Flow (cubic feet per second)}}{\text{Runner Outside Diameter (inches)} \times \sqrt{\text{Head (feet)}}}$$

$$B_2 = \text{Runner Width between Discs} = B_1 + 1/2 \text{ to } 1"$$

$$\text{Rotational Speed (revolutions per minute)} = \frac{73.1 \times \sqrt{\text{Head (feet)}}}{\text{Runner Outside Diameter (feet)}}$$

The efficiency of the Michell turbine is 80% or greater and therefore suitable for small power installations. Flow regulation and governor control of the flow can be effected by using a center-body nozzle regulator (a closing mechanism in the shape of a gate in the nozzle). This is expensive because of governor costs. It is, however, needed for running an alternating-current generator.

The application of Figures 25 and 26 is a typical example. For high heads the Michell turbine is connected to a penstock with a turbine inlet valve. This requires a different type of arrangement from the one shown here. As mentioned before, the Michell turbine is unique because its B_1 and B_2 widths can be altered to suit power-site traits of flow rate and head. This, besides simplicity and low cost, makes it the most suitable of all water turbines for small power developments.

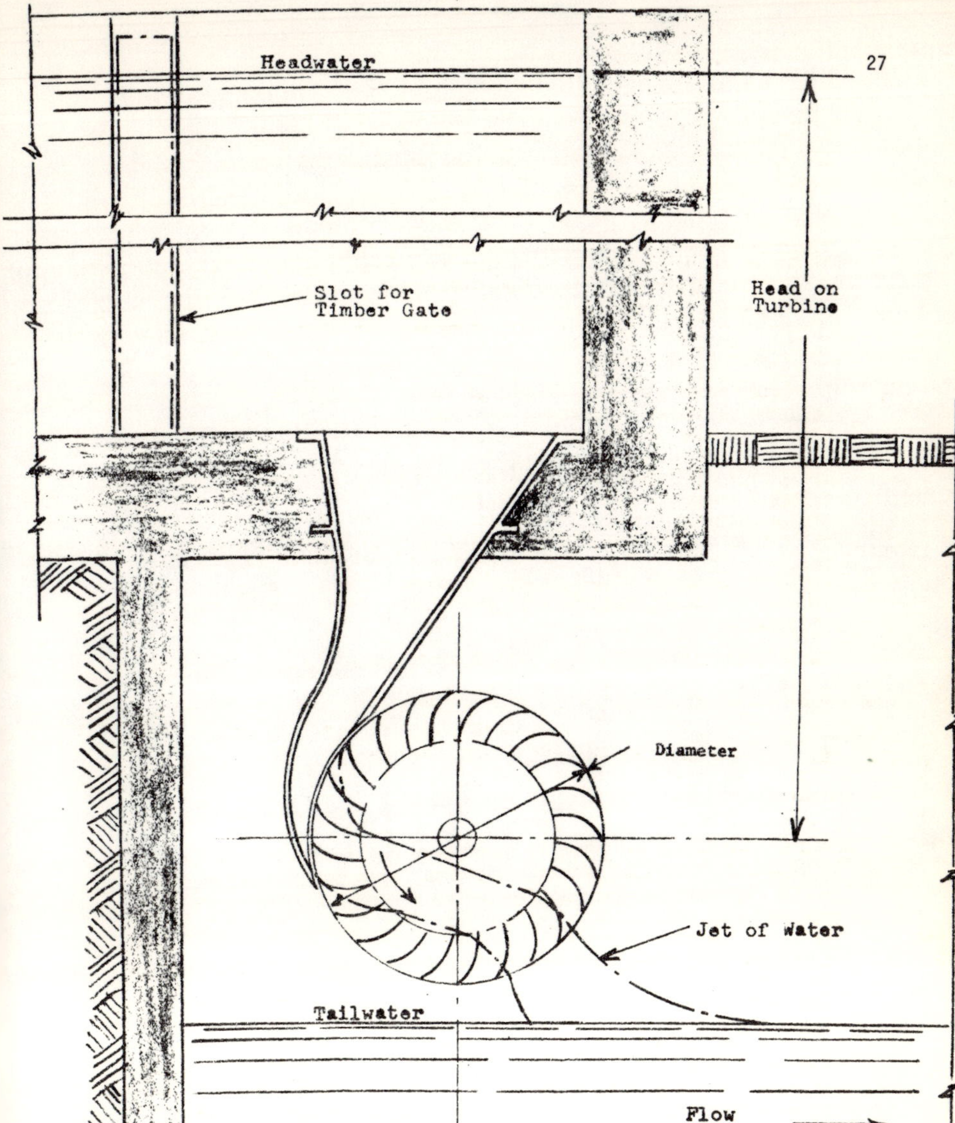

Figure 25. Arrangement of a Michell (Banki) turbine for low-head use without control (A).

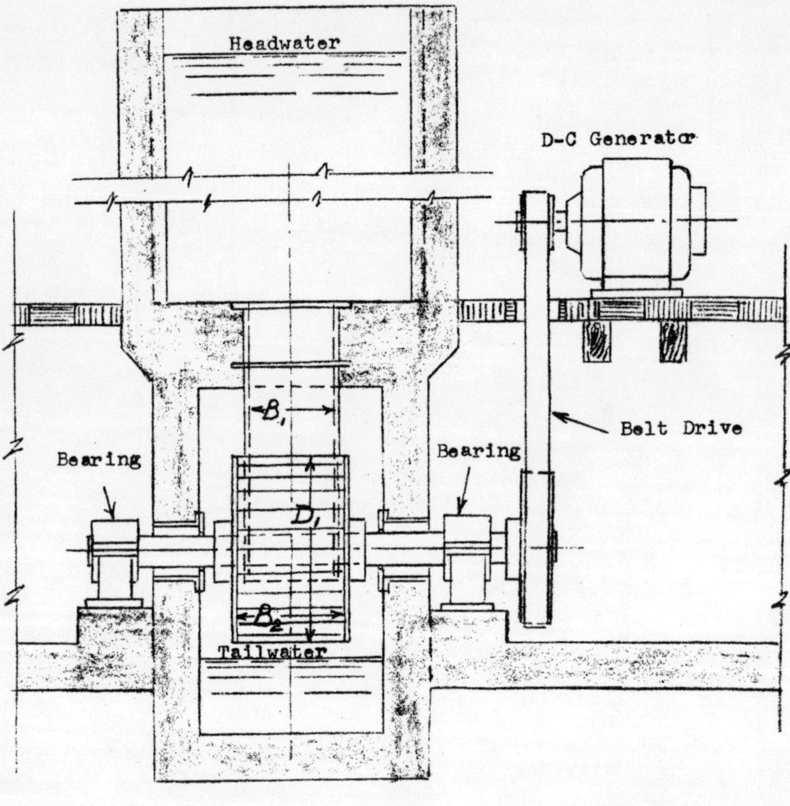

Figure 26. Arrangement of a Michell (Banki) turbine for low-head use without control (B)

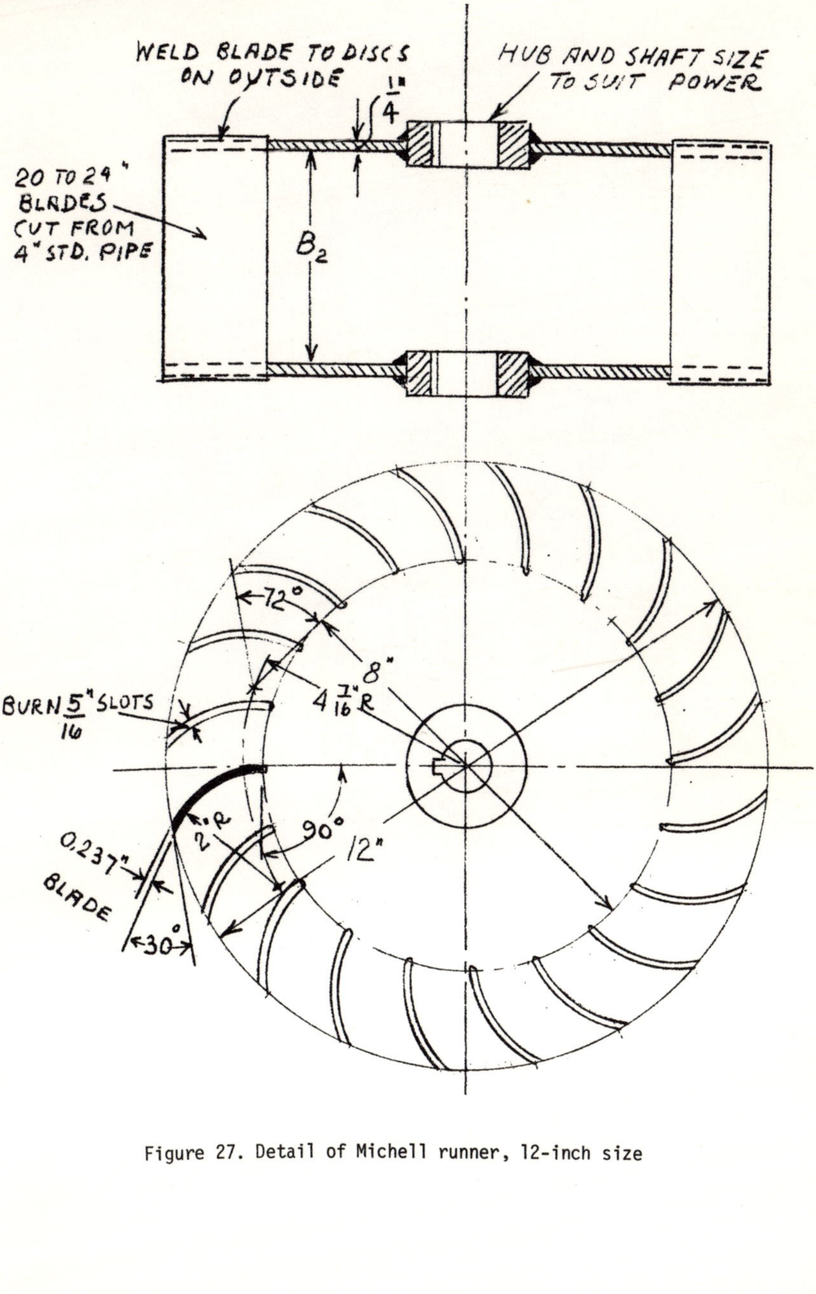

Figure 27. Detail of Michell runner, 12-inch size

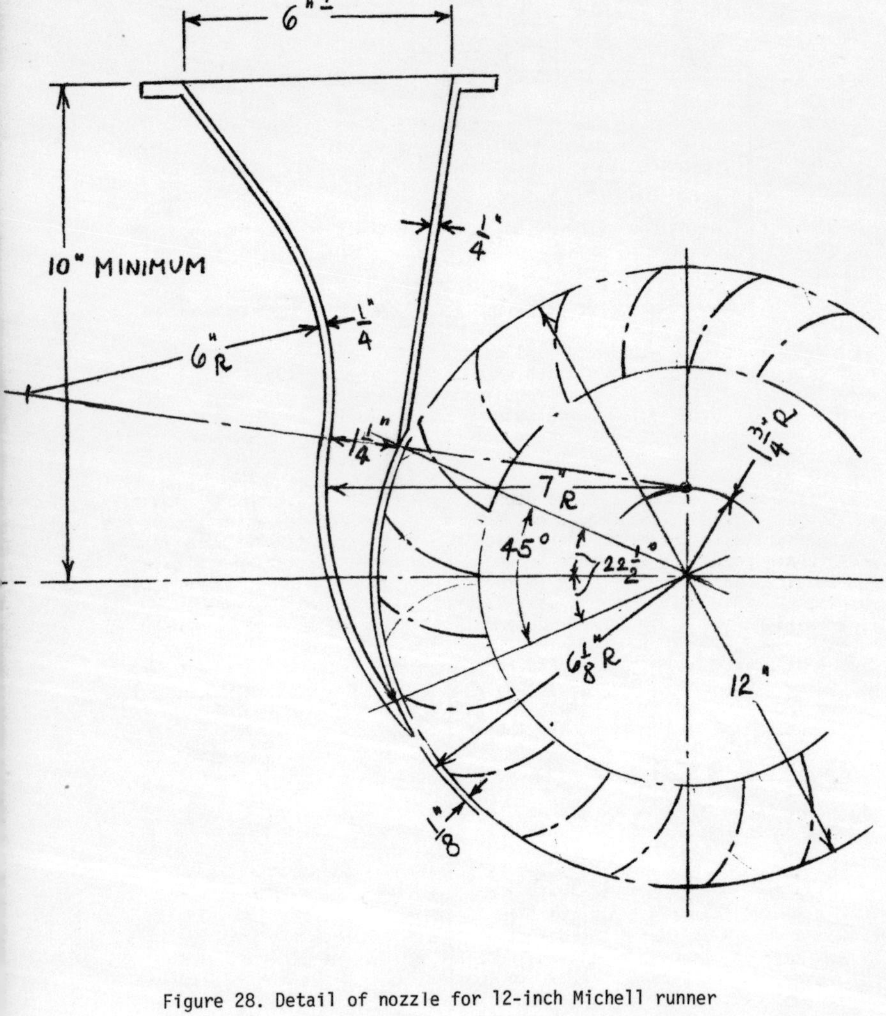

Figure 28. Detail of nozzle for 12-inch Michell runner

C. Centrifugal Pumps and Propeller-Type Pumps

The use of centrifugal pumps or propeller-Type pumps as turbines should be explored before all other alternatives, provided that direct-current electricity can be used (See Figures 29 and 30). Centrifugal pumps and propeller-type pumps are relatively low in cost and are available in many sizes. Manufacturers can quote the proper unit if head and flow are given.

They can be used to produce alternating current also, but with increased cost. In this case, a butterfly valve is used as the turbine-inlet valve; and the valve can be regulated by a small water-turbine governor.

The help of an engineer should be sought in modifying these pumps for use as turbines.

IX. WATER WHEELS

Water wheels date back to biblical times but are far from obsolete. They have certain advantages which should not be overlooked. They are more economical for small power requirements than water turbines in some cases. It is possible to make a water wheel for power requirements up to 10 horsepower in places where there are no elaborate manufacturing facilities.

Water wheels are attractive especially where fluctuations in flow rate are large. Speed regulation is not practical - therefore, water wheels are used primarily to drive machinery which can take large fluctuations in rotational speed. They operate between 2 and 12 revolutions per minute and require gearing and belting (with inherent friction loss) to run most machines. Thus, they are most useful for slow-speed applications, e.g., flour mills, some agricultural equipment, and some pumping operations.

A water wheel, because of its rugged design, requires less care than a turbine does. It is self-cleaning, and, therefore, need not be protected from debris (leaves, grass and stones). The two main types of water wheels are the overshot and the undershot.

A. Overshot Water Wheel

The overshot water wheel may be used with heads of 10 to 30 feet, and flow rates from one to 30 cubic feet per second.

The water is guided to the wheel in a timber or metal flume at a water velocity of approximately 3 feet per second. A gate at the end of the flume controls the flow to the wheel and the jet velocity, which should be from 6 to 10 feet per second. To obtain this velocity, the head (H_1 in Figure 31) should be one to two feet. Wheel width depends on the amount of water to be used. The discharge will be one to two cubic feet per second for a flume width of one foot. Wheel width must exceed flume width by about one foot because of jet expansion. The efficiency of a well-constructed overshot water wheel can be 60% to 80%.

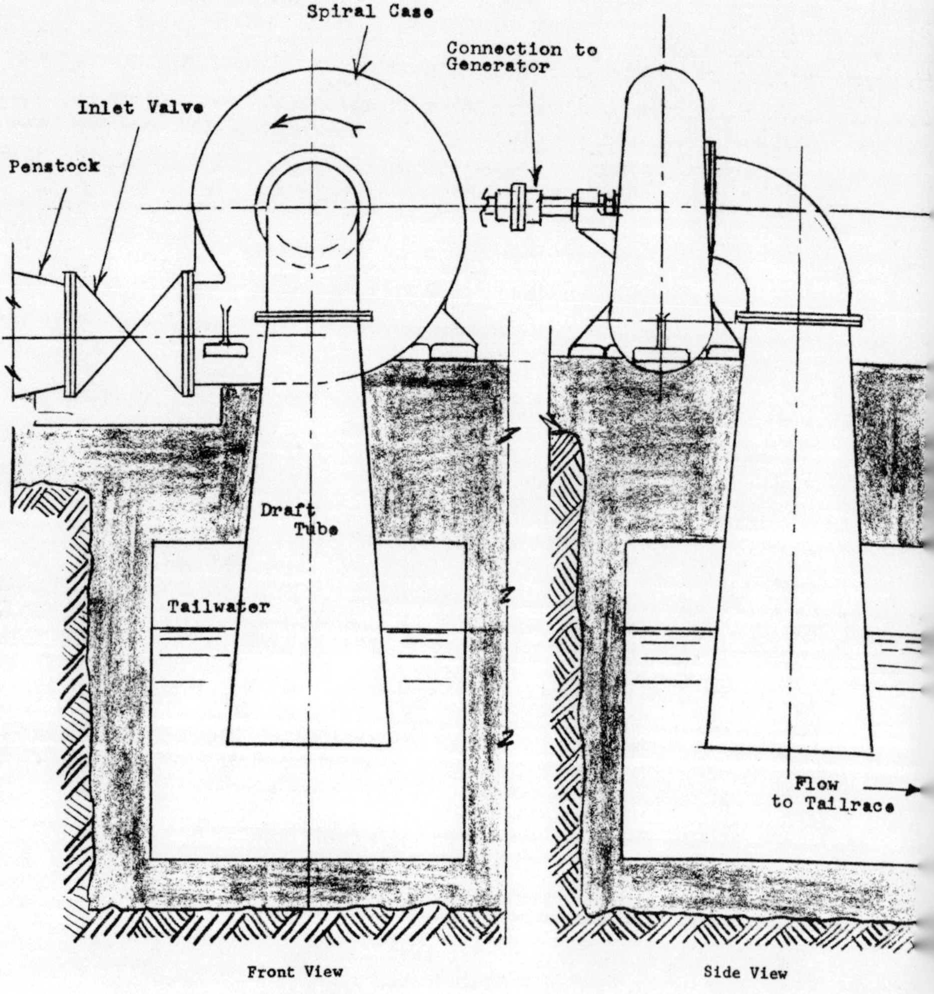

Figure 29. Centrifugal Pump which, with reversed rotation, can function as a Water Turbine

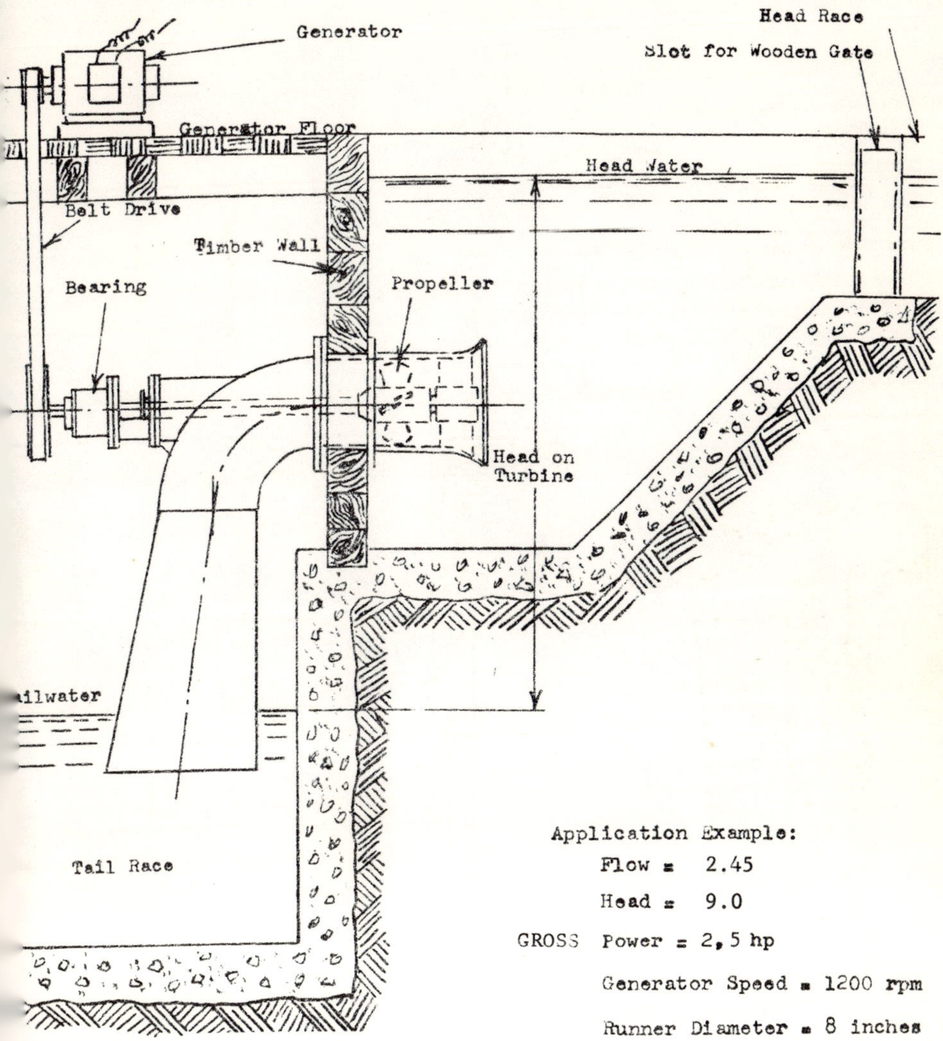

Figure 30. Simple Propeller Pump which, with reversed rotation, can function as Propeller Turbine

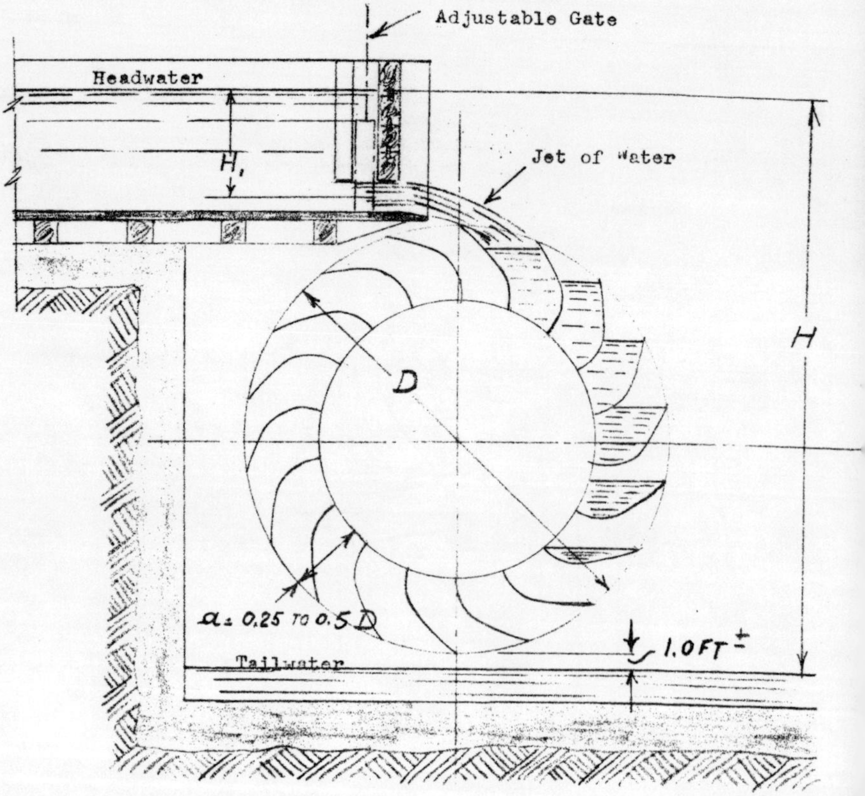

Figure 31. Overshot Water Wheel

B. Undershot Water Wheel

The undershot water wheel (Figure 32) should be used with heads of 1.5 to 10 feet and flow rates from 10 to 100 cubic feet per second. Wheel diameter should be 3 to 4 times the head -- wheel diameters between 6 and 30 feet. Rotational speed should be 2 to 12 revolutions per minute, with the higher speed applying to the smaller wheels. For each foot of wheel width, the flow rate should be between 3 and 10 cubic feet per second. The wheel dips from one to three feet into the water. Efficiency is in the range of 60% to 75%.

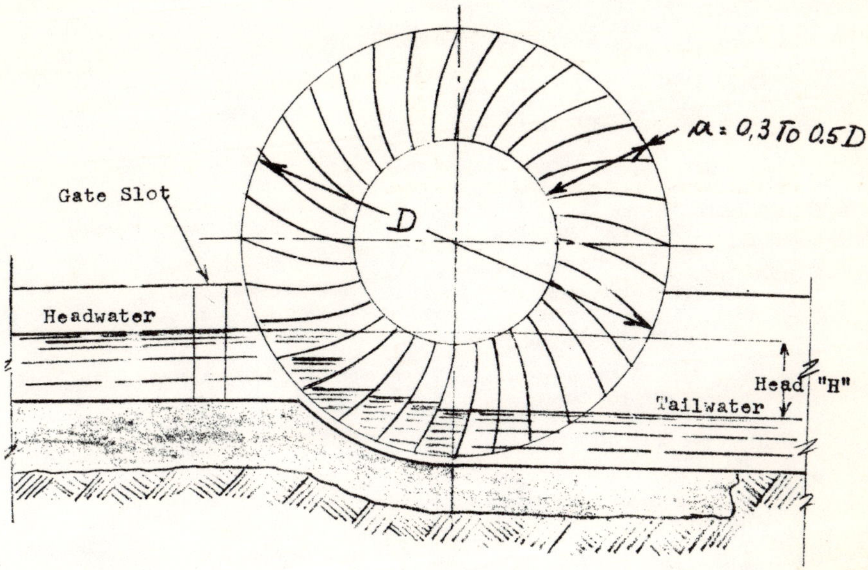

Figure 32. Undershot Water Wheel

X. EXAMPLES

<u>Mission Hospital</u>

1. Requirements: 10 kilowatt light and power plant.
2. 10 kilowatts is 13 1/3 horsepower.
3. The gross power required is then about 27 horsepower.
4. A stream in hilly territory can be dammed up and the water channeled through a ditch 1/2 mile long to the power plant site.
5. A penstock 250 feet long will take the water to the turbine.
6. The total difference in elevation is 140 feet.
7. Available minimum flow rate: 1.8 cubic feet/second.

8. The soil in which the ditch is to be dug permits a water velocity of 1.2 feet per second.

9. Table II, Section VI gives n= 0.030

10. Area of flow in the ditch = 1.8/1.2 = 1.5 square feet.

11. Bottom width = 1.5 feet.

12. Hydraulic radius = 0.31 X 1.5 = 0.46 feet.

13. Figure 8 shows that this results in a fall and head loss of 1.7 feet for 1,000 feet. The total for the half-mile (2,640 feet) ditch is 4.5 feet.

14. The fall that is left through the penstock is then: 140-4.5 = 135.5 feet. Figure 10 gives 5.7 inches as the required penstock diameter for 1.8 cubic feet per second flow at 10 feet per second velocity.

15. Head loss in the penstock is 10 feet for 100 feet of length and 25 feet for the total length of 250 feet.

16. For the water tubine:
Net Head = 135.5-25 = 110.5 feet

17. Power produced by the turbine at 80% efficiency:

$$\text{Net Power} = \frac{\text{Minimum water flow X net head}}{8.8} \text{ X Turbine Efficiency}$$

$$= \frac{1.8 \text{ X } 110.5}{8.8} \text{ X } .80 = 18 \text{ horsepower}$$

18. Consult Table III. The cost of a pump or turbine for a particular situation can only be learned by writing to the various manufacturers. VITA engineers can step in here, lay out the physical arrangement and compile a list of necessary mechanical and electrical components to the best advantage of the field worker.

Appendix 1

AVAILABILITY OF MANUFACTURED TURBINES

Small hydraulic turbines and even more the governors for regulating these turbines are difficult to obtain because the demand for these products has diminished to a considerable extent in the last twenty years. And manufactured water wheels are completely off the market. Of the remaining number of manufacturers of small turbines and governors only one exists in the United States, and two are known by the author to exist in Europe.

The James Leffel & Company is located in Springfield, Ohio. Their booklet, Leffel Pamphlet "A", Hints on the Development of Small Water Power, is available on request. It is a very useful supplement to the information in this manual. Its description of Leffel's small vertical Samson turbine is very complete. This turbine is available in sizes from 3 to 29 horsepower. The company maintains an engineering department which stands ready to assist in planning and design of the entire installation.

This company also manufactures a complete unit called Hoppes Hydroelectric Unit, which is useful in isolated locations where the demand is small. It comes in sizes of 1 to 10 kilowatts. A Leffel bulletin describing this unit gives complete instructions on submitting the information necessary for ordering it.

The Michell (or Banki) turbine is manufactured exclusively by the Ossberger-Turbinenfabrik of Weissenburg, Bavaria, Germany. This turbine is made in sizes ranging from 1 to 1000 horsepower. The company has an impressive record of installations, many in less-developed countries. Ossberger-Turbinenfabrik is very responsive to requests for information. It furnishes without charge a considerable amount of data, translated into English. The simple design of the Michell turbine makes it a favorite for remote regions and is priced lower than corresponding Francis and impulse type turbines. Its governor, developed by Ossberger, is also very reasonably priced.

A third company which manufactures turbines and governors for turbines but does not sell packaged units, including the electrical equipment, is the Officine Buehler, Taverne, Canton Ticino, Switzerland. They are in the small turbine field, and they manufacture all types except Michell. Their workmanship is of the highest quality, and their engineering is superb. Like the other companies, they assist prospective customers in planning their installations.

Appendix 2

CONVERSION TABLES

Units of Length

1 Mile	= 1760 Yards	= 5280 Feet
1 Kilometer	= 1000 Meters	= 0.6214 Mile
1 Mile	= 1.607 Kilometers	
1 Foot	= 0.3048 Meter	
1 Meter	= 3.2808 Feet	= 39.37 Inches
1 Inch	= 2.54 Centimeters	
1 Centimeter	= 0.3937 Inch	

Units of Area

1 Square Mile	= 640 Acres	= 2.5899 Square Kilometers
1 Square Kilometer	= 1,000,000 Sq. Meters	= 0.3861 Square Mile
1 Acre	= 43,560 Square Feet	
1 Square Foot	= 144 Square Inches	= 0.0929 Square Meter
1 Square Inch	= 6.452 Square Centimeters	
1 Square Meter	= 10.764 Square Feet	
1 Square Centimeter	= 0.155 Square Inch	

Units of Volume

1.0 Cubic Foot	= 1728 Cubic Inches	= 7.48 U.S. Gallons
1.0 British Imperial Gallon	= 1.2 U.S. Gallons	
1.0 Cubic Meter	= 35.314 Cubic Feet	= 264.2 U.S. Gallons
1.0 Liter	= 1000 Cubic Centimeters	= 0.2642 U.S. Gallons

Units of Weight

1.0 Metric Ton	= 1000 Kilograms	= 2204.6 Pounds
1.0 Kilogram	= 1000 Grams	= 2.2046 Pounds
1.0 Short Ton	= 2000 Pounds	

CONVERSION TABLES

Units of Pressure

1.0 Pound per square inch	= 144 Pound per square foot
1.0 Pound per square inch	= 27.7 Inches of Water*
1.0 Pound per square inch	= 2.31 Feet of Water*
1.0 Pound per square inch	= 2.042 Inches of Mercury*
1.0 Atmosphere	= 14.7 Pounds per square inch (PSI)
1.0 Atmosphere	= 33.95 Feet of Water*
1.0 Foot of Water = 0.433 PSI	= 62.355 Pounds per square foot
1.0 Kilogram per square centimeter	= 14.223 pounds per square inch
1.0 pound per square inch	= 0.0703 kilogram per square centimeter

* at 62 degrees Fahrenheit (16.6 degrees Celsius)

Units of Power

1.0 Horsepower (English)	= 746 Watt = 0.746 Kilowatt (KW)
1.0 Horsepower (English)	= 550 foot pounds per second
1.0 Horsepower (English)	= 33,000 foot pounds per minute
1.0 Kilowatt (KW) = 1000 Watt	= 1.34 Horsepower (HP) English
1.0 Horsepower (English)	= 1.0139 Metric Horsepower (cheval-vapeur)
1.0 Metric Horsepower	= 75 Meter X Kilogram/Second
1.0 Metric Horsepower	= 0.736 Kilowatt = 736 Watt

Appendix 3

BIBLIOGRAPHY

General Texts and Handbooks

Brown, J. Guthrie ed, Hydro Electric Engineering Practice. New York: Gordon & Breach, 1958; London: Blackie and Sons, Ltd., 1958. A very complete treatise covering the entire field of hydroelectric engineering. Three volumes. v. 1 Civil Engineering $50.00 U.S. v. 2 Mechanical and Electrical Engineering $30.00 U.S. v. 3 Economics, Operation and Maintenance ($25.00 U.S.)

Creager, W. P. and Justin, J. D. Hydro Electric Handbook. 2d ed. New York: John Wiley and Son, 1950. A most complete handbook covering the entire field. Especially good for reference. ($18.50 U.S.)

Davis, Calvin V. Handbook of Applied Hydraulics. 2d ed. New York: McGraw-Hill, 1952. A comprehensive handbook covering all phases of applied hydraulics. Several chapters are devoted to hydroelectric application. ($23.50 U.S.)

Paton, T. A. L. Power from Water. London: Leonard Hill, 1961. A concise general survey of hydroelectric practice in abridged form. ($8.50 U.S.)

Zerban, A. H. and Nye, E.P. Power Plants. 2d ed. Scranton, Penn.: International Text Book Co., 1952. Chapter 12 gives a concise presentation of hydraulic power plants. ($8.00 U.S.)

The Banki Turbine

Haimerl, L. A., "The Cross Flow Turbine," Water Power (London), January 1960. Reprints available from Ossberger Turbinenfabrik, 8832 Weissenburg, Bayern, Germany. This article describes a type of water turbine which is being used extensively in small power stations, especially in Germany.

Mockmore, C. A. and Merryfield, F., The Banki Water Turbine. Corvallis, Ore.: Oregon State College Engineering Experiment Station Bulletin No. 25, February 1949. 40¢. A translation of a paper by Donat Banki. A highly technical description of this turbine, originally invented by Michell, together with the results of tests.

Appendix 4

THE AUTHOR AND REVIEWERS

Hans W. Hamm, a VITA Volunteer, was a consultant on small water power developments for twenty years with a Pennsylvania manufacturer of water wheels and small turbines. He earned a degree in mechanical engineering from the State Technical University of Braunschweig in his native Germany. He retired in 1966 from the York, Pennsylvania, works of Allis-Chalmers.

***

Other VITA Volunteers have helped in producing this manual: Morton Rosenstein, public relations and market research manager at Ionics, Inc., Watertown, Massachusetts, edited the entire manual.

Harry Wiersema, consulting engineer of Knoxville, Tennessee, made many helpful suggestions based on more than fifty years' experience in hydraulic engineering. He also wrote the preface for the manual and prepared the bibliography.

Dr. John J. Cassidy, associate professor of civil engineering, University of Missouri, and Robert H. Emerick, consulting engineer of Charleston, South Carolina, both reviewed the manual for technical accuracy.

Ian D. Burnet, projects officer of the Department of Trade and Industry, Port Moresby, Papua, New Guinea, reviewed the book from the point of view of the eventual user, the community development leader.

***

The manual was also reviewed by Jeffrey Ashe and John Brandi, Peace Corps Volunteers who were working on a project to develop a small water power site in Loja, Ecuador, by Ossberger Turbinenfabrik, Weissenburg (Bayern), Germany and by James Leffel & Company, Springfield, Ohio.

Appendix 5

DATA SHEET

This form is given as a guide to help you collect the information a VITA engineer would need to help you plan a small water power site.

TO: VITA
3706 Rhode Island Avenue
Mt. Rainier, Maryland 20822 USA

1. Minimum flow of water available in cubic feet per second (or cubic meters) per second. ______________
2. Maximum flow of water available in cubic feet per second (or cubic meters) per second. ______________
3. Head or fall of water in feet (or meters) ______________
4. Length of pipe line in feet (or meters) needed to get the required head. ______________
5. Describe water condition (clear, muddy, sandy, acid) ______________
6. Describe soil condition (see Table II) ______________
7. Minimum tailwater elevation in feet (or meters)______________
8. Approximate area of pond above dam in acres (or square kilometers). ______________
9. Approximate depth of the pond in feet (or meters)______________
10. Distance from power plant to where electricity will be used in feet (or meters). ______________
11. Approximate distance from dam to power plant ______________
12. Minimum air temperature. ______________
13. Maximum air temperature. ______________
14. Estimate power to be used. ______________
15. ATTACH SITE SKETCH WITH ELEVATIONS, OR TOPOGRAPHICAL MAP WITH SITE SKETCHED IN.

Date______________

Name ______________
Address ______________
____
____

See reverse for guide in collecting further helpful information.

DATA SHEET - 2

The following questions cover information which, although not necessary in starting to plan a water power site, will usually be needed later. If it can possibly be given early in the project, this will save time later.

1. Give the type, power and speed of the machinery to be driven and indicate whether direct, belt or gear drive is desired or acceptable.

2. For electric current, indicate whether direct current is acceptable or alternating current is required. Give the desired voltage, number of phases and frequency.

3. Say whether manual flow regulation can be used (with D.C. and very small A.C. plants) or if regulation by an automatic governor is needed.

VITA Bookshop

Weave your own fabric with...

Handloom Construction

by Joan Koster

Beautifully illustrated with line drawings, this easy-to-follow guide shows you how to build three different kinds of handlooms that you can make and use in your own home. Guidelines are provided for judging which loom to build, and the best materials to use for weaving. Expert tips are included on weavers' tools, patterns, and finishing touches for professional results. A great introduction to a fascinating pastime and hobby. Plastic bound for easy reference. 164 pages, 8-1/2" x 11", softcover.

39 Handloom Construction $8.95

New printing now in stock!

Village Technology Handbook

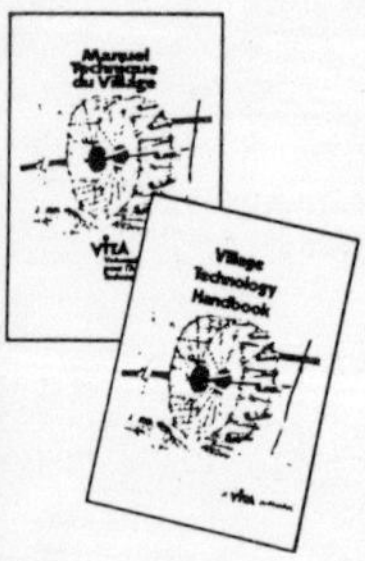

A must for development workers and do-it-yourselfers, this practical manual of small-scale technology offers expert advice on everything from water power to village crafts and industries. Fully illustrated with photographs, graphs, charts, and line drawings, its easy-to-use construction plans cover well digging, food storage, sanitation, hand-powered washing machines, and much more. A handy book for homesteaders and country dwellers.
390 pages, 5-1/2" x 8", softcover.
Also available in French and Spanish editions.

1E Village Technology Handbook (English Edition)........$9.95

1F Village Technology Handbook (French Edition)........$10.95

1S Village Technology Handbook (Spanish Edition).......$10.95

HYDRAULIC RAM FOR TROPICAL CLIMATES

by Allen Inversin

Construct a low-cost ram from commercially availabl pipe fillings. Two homemade valves require only a dril press and handtools to make. This pump will deliver u to several thousand liters of water per day and require very little maintenance. Fully illustrated with step-by step instructions. 1979, 50 pages, 5½" x 8½", softcover ISBN 0-86619-070-8

11050 Hydraulic Ram for Tropical Climates $3.9

OVERSHOT WATER-WHEEL: DESIGN AND CONSTRUCTION MANUAL

VITA

This low-cost water-wheel provides power to pump water, grind grain, press oils or sugar cane. Constructio details are provided, and design considerations are give for both small and large overshot water-wheels. 1979, 74 pages, 8½" x 11", softcover. ISBN 0-86619-067-8

11047 . Overshot Water-Wheel $5.9

SMALL MICHELL (BANKI) TURBINE: A CONSTRUCTION MANUAL

VITA

A low-cost water turbine that can provide ac dc electricity for your home. Complete step-by-step instructions fo making parts and assembly. Illustrated. 1979, 58 pages, 8½" x 11", softcover. ISBN 0-86619-066-X

11045 Small Michell (Banki) Turbine $6.9

From China!

Biogas Blueprints

A complete set of blueprints of the Fixed Dome Biogas Digester is now available from VITA. Used throughout the Peoples Republic of China, this digester has proven to be a cost-effective way of generating methane gas, and improving sanitation. To order your complete set of 25 blueprints, just fill out the coupon on the facing page. The set is only $30.00, postpaid. Only available in English.

Fixed Dome Biogas Digester Blueprints
01005 ...$30.00

ABOUT VITA

Volunteers in Technical Assistance (VITA) is a private, non-profit, international development organization. It makes available to individuals and groups in developing countries a variety of information and technical resources aimed at fostering self-sufficiency--needs assessment and program development support; by-mail and on-site consulting services; information systems training.

VITA promotes the use of appropriate small-scale technologies, especially in the area of renewable energy. VITA's extensive documentation center and worldwide roster of volunteer technical experts enable it to respond to thousands of technical inquiries each year. It also publishes a quarterly newsletter and a variety of technical manuals and bulletins.

VITA's documentation center is the storehouse for over 40,000 documents related almost exclusively to small- and medium-scall technologies in subjects from agriculture to wind power. This wealth of information has been gathered for almost 20 years as VITA has worked to answer inquiries for technical information from people in the developing world. Many of the documents contained in the Center were developed by VITA's network of technical experts in response to specific inquiries; much of the information is not available elsewhere. For this reason, VITA wishes to make this information available to the public.

For more information, contact VITA at 3706 Rhode Island Avenue, Mount Rainier, Maryland 20822 U.S.A.